Learning to Fly Gliders

A Flight Training Handbook and Syllabus

Bob Wander

ISBN 0-9636227-0-6

Fourth Edition. Published By

SOARING BOOKS & SUPPLIES

P.O. Box 17250 Minneapolis, MN 55417

(800) 660-0238 (952) 920-1804

Email: SoarBooks@aol.com

Online Ordering at www.bobwander.com

Bob Wander's *Gliding...Made Easy!* Series

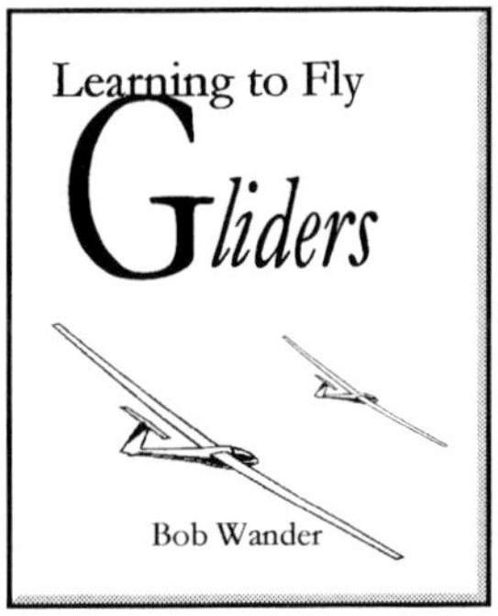

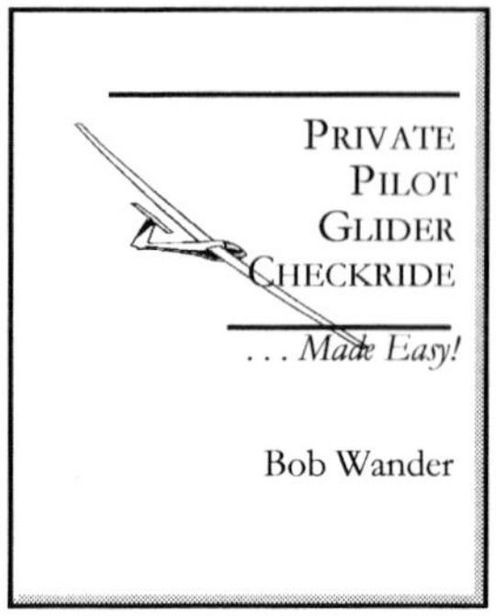

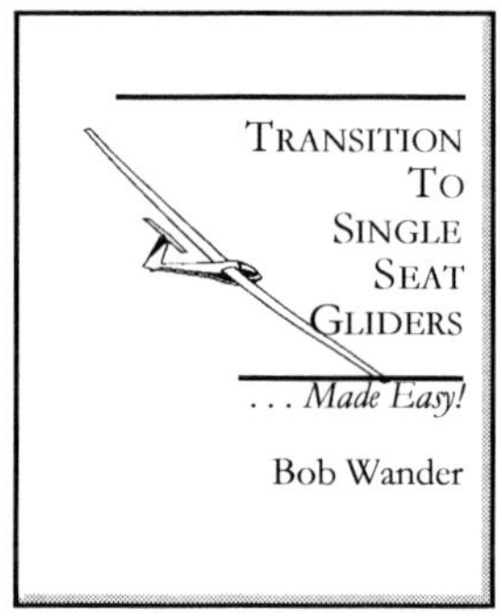

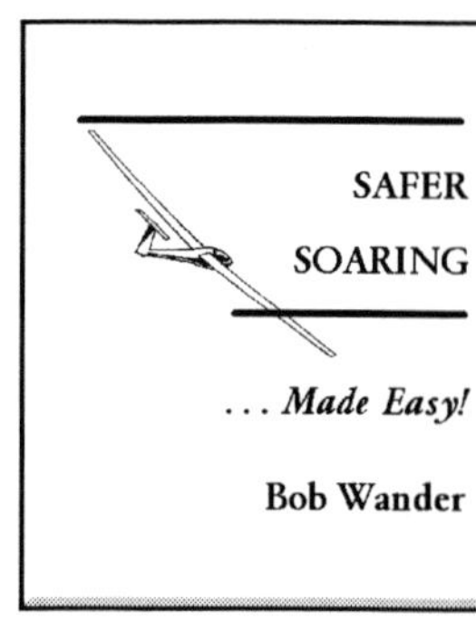

Soaring's Best Sellers! Available From:

SOARING BOOKS & SUPPLIES

(800) 660-0238 (952) 920-1804

P.O. Box 17250 Minneapolis, MN 55417

Purchase Online at www.bobwander.com

Introduction

This glider flight-training manual provides you with a clear picture of the decision-making skills and flight maneuvers you must master to become a competent soaring pilot.

Learning to fly well is a big job. As with many big jobs, the best way to accomplish it is to break it down into a series of little jobs, so that you can concentrate on one job at a time. This book does that for you. I have organized the book so that each section treats a specific area of flight training. Within each section, not only is a big job divided into lots of little jobs, but the little jobs are presented in progressive order. Structuring each section in this way provides you with a step-by-step approach to building your flying skills. It also affords you the opportunity to look ahead and prepare for your next lesson before you get into the glider.

In each lesson, you learn the objective, why the objective is important for you to learn, and how your instructor will present the material to you. Most lessons also list common student errors, so that you will be forewarned against them... as well as not too discouraged if you find yourself temporarily frustrated by them. Finally, each lesson contains a descriptive completion standard, so that you can measure and record your progress, and take justifiable pride in your accomplishments.

When I was learning to fly gliders, no book like this was available. I often wished for one. I hope this book helps you make your dream of motorless flight come true.

Bob Wander

Table of contents

How to use this book

Glider flight instruction depends a great deal on the weather. Some days may be too windy or rainy for you to fly. Others may not be windy enough, as seems to be the case whenever I have a student ready to tackle crosswind landings — the crosswinds disappear for a week at a time. A good training program for glider pilots cannot ignore these real-world considerations. They have not been ignored in this book.

When you use this book, it is best to consider the material within each section as sequential lesson plans. For instance, in Section 2, you are introduced to the basic motions of the flight controls before you are asked to perform shallow turns and straight glides. In Section 3 you learn to fly the aerotow in high tow position before you learn to fly the glider all around the wake of the towplane in a wake box maneuver.

That the material within each section is sequential is no accident, and is intended to break down the large job of learning to fly gliders into a number of smaller, more manageable jobs that you can learn one at a time. It is not necessary, however, to consider the sections themselves as strictly sequential. For instance, Section 3 is concerned with takeoffs and aerotows, while Section 4 is concerned with patterns and landings. It is quite normal for a student to work on aerotow and on pattern and landing in the same flight (although not, I hope, at the same instant!). In this case, you may have progressed to the middle of Section 3 (which is about aerotows), and may have progressed a third of the way through

Section 4 (which is about patterns and landings). If you are ready on a given day for crosswind patterns and landings, and if on that day no crosswind can be had, there is no problem — you and your instructor can pick up in Section 3 (aerotows) where you left off, and once off tow you can progress in Section 5 (advanced free-flight maneuvers) from where you left off there. You are free to make progress in several areas. This freedom to move around a bit among sections in your training program allows you to make the most of your flight time. Progress record sheets are provided at the end of the book for you and your instructor to keep track of your progress.

Finally, a note about the scope of this book. It is intended to be used as a guide to effective flight training. As such, it is a supplement to, not a replacement for, a thorough program of study of all the topics with which a glider pilot must be familiar. Aerodynamics, weather, airspace utilization, aviation regulations, and a small host of other topics must be studied and understood before you can become a safe pilot. An excellent book designed to assist you in studying these topics is the *Soaring Flight Manual*, published by the Soaring Society of America. I recommend it highly and I use it for ground instruction with all my students. Your instructor will guide your ground study. A bibliography of useful books is included in the Appendix.

Notes to instructors

When I think back and remember all the teachers I have ever had in my lifetime, I have to conclude that only a handful of them were truly outstanding educators. Before I began to write this book, I thought long and hard about what made them such excellent teachers. When I was satisfied that I had identified those qualities, I set out to incorporate them into this book.

Learning To Fly Gliders offers you a pragmatic and effective flight training program. Ease of use, for you and for your students, was always on my mind as I wrote it and shaped it. The writing style is uncomplicated and to the point. Lessons are sensibly grouped by the phases of flight—preflight and ground handling, launching, free flight, and landing. Each lesson plan provides you with a lesson (what to teach), a purpose (why it is important for the student to learn), a procedure (suggestions on how to teach it), common errors (a trouble-shooting guide to the student's learning process), and a completion standard (the goal). Once the completion standard is attained, it is noted in the progress record, which both documents past progress and helps you select future lessons. When combined with the *Soaring Flight Manual* (an excellent ground school study guide by the Soaring Society of America), *Learning To Fly Gliders* provides your students with a comprehensive soaring education program.

Whether your glider students are brand new to flying or have 10,000 hours in airplanes, you can use this book with them. The FAA *Private Glider Pilot Practical Test Standards* holds each glider applicant to the same knowledge and performance standards. For that reason, it makes sense to train each student using the same set of completion standards. Naturally, most power pilots will progress rapidly through the early lessons, but launching, thermaling, advanced airwork, soaring decision-making, and traffic patterns will present many new challenges.

Learning To Fly Gliders leaves you lots of breathing room, so that you can develop and employ your own style of flight instruction. All decisions about first solo, length of lessons, and so on are left in your hands. Only you can make those decisions. I encourage you to incorporate your own experience and background into your employment of this book, and to modify its content as your judgment and flying environment demand.

Pre-flight and ground handling

Section 1

The lessons in this section help you to learn many important features and characteristics of your glider. Instrument systems, assembly and disassembly procedures, ground handling and transport, and pre-flight inspection may not be glamorous topics, but they are very important to the safety of flight and to the proper care and safe operation of the glider. Make sure that you procure a copy of the pilot's manual for the type of glider you will be training in. Learn as much as you can about the systems, performance characteristics, ground handling, and assembly and disassembly procedures of your training glider.

The experience of extracting all possible pertinent information from the operating manual of your training glider will stand you in good stead later in your gliding career, when you will have the opportunity to learn about and fly other types of gliders.

Glider familiarization

Lesson

Glider, glider cockpit familiarization

Glider manual

Ground handling

AROW (Airworthiness certificate, Registration, Operating limitations, Weight-and-balance limitations) aboard

Glider weight and balance, ballast (if applicable)

Performance speeds

Purpose

This lesson familiarizes you with the glider, including an introduction to its inspection, airworthiness requirements, FAA required paperwork, loading limits, and critical performance speeds.

Procedure

With your instructor you will examine the glider rigging and flight manual, aircraft logbook, airworthiness and registration certificates, and radio station license (if the glider is radio-equipped). You and your instructor will discuss the glider's operating limitations, including the V-speeds and their importance, and the weight-and-balance limitations of the glider and their effect on flight characteristics. The use of forward ballast, if required, also will be discussed.

Common errors

Failure to procure and study the aircraft manual

Completion standard

This lesson is complete when you can:

- Describe the basic requirements that the glider must meet to be legal for flight
- Explain stall speed (clean), stall speed (dirty), minimum sinking speed, best L/D speed, maneuvering speed, rough air speed, and never-exceed speed
- Explain the importance of operating the glider within approved weight-and-balance limits

Pre-flight, ground handling, airport familiarization

Lesson

Glider pre-flight and ground handling

Familiarization with the taxiways and runways of your airfield

Purpose

In this lesson you learn how to determine if the glider is ready for flight. You will then learn how to move the glider to the flightline, and learn the locations of the runways and taxiways of your airfield.

Procedure

Using the manufacturer's aircraft manual, your instructor will guide you through a pre-flight inspection of the glider. Pre-flight inspection of the primary flight controls will include a positive control check, which your instructor will perform with you. When the pre-flight is complete and the glider is in safe condition for flight, you will learn how to move the glider to the flightline using a tow vehicle and a long rope. Your instructor will also explain the routes and conventions at your home airfield for moving the glider along taxiways and runways to the flightline.

Common errors

Failure to use the aircraft manual checklist for pre-flight inspection

Failure to perform a positive control check

Ground towrope too short

Failure to release the glider from the tow vehicle on arrival at the flightline

Wrenching the glider around turns using the wingtips as levers

Failure to ensure that both wingtips will pass between obstacles or restrictions

Towing too fast

Failure to secure the control stick

Failure to have the wing-runner make sure the traffic pattern is clear of conflicting traffic before crossing an active runway

Completion standard

This lesson is complete when you can perform a thorough pre-flight inspection, supervise ground tow to the flightline in accordance with procedures in use at your airport, and on arrival at the flightline, disconnect the glider from the tow vehicle and secure the glider.

Flight instruments

Lesson Introduction to the flight instruments

Purpose This lesson introduces the airspeed indicator, the altimeter, the compass, the variometer, and the yaw string.

Procedure The **airspeed indicator** does not actually sense airspeed. Rather, it senses two pressures, compares them, and displays the difference between the two. If pitot (or ram air) pressure is greater than ambient pressure, the airspeed indicator displays a positive airspeed value. If the pressures are the same, the indicator reads zero. The airspeed indicator confirms visually what pitch attitude, control feel, and wind noise already tell us about the speed of the glider. Due to changes in atmospheric pressure and density, airspeed indicators are more and more inaccurate as altitude increases, indicating airspeeds that are lower and lower with respect to true airspeed. Note that if at any altitude the pitot tube or static ports are blocked by water, foreign matter, protective covers, or insect debris, the airspeed indicator function will be compromised; its indication of airspeed will be inaccurate. The other cues to airspeed mentioned above, however, will still be reliable.

The **altimeter** in the glider does not actually sense altitude. What it does sense is the effect of ambient pressure on the dimensions of a sealed aneroid bellows. If pressure is lowered, the bellows swells like a little accordion, and the altimeter shows an increase in altitude. If pressure rises, the little accordion is compressed, and the altimeter shows a decrease in altitude. Since the altimeter senses pressure, not altitude, it should be used as an approximate guide to altitude — no more than that. In particular, in the approach to landing the altimeter should be used only to confirm what you have already learned by keeping your eyes outside of the cockpit as you judge the angles to your aiming spot during your approach — that you have sufficient height to make a safe approach and landing. Note that atmospheric pressure (barometric pressure) can vary widely due to the weather. As a result, the altimeter has a small knob that allows you to adjust for changes in barometric pressure.

The **compass**, as installed in an aircraft, does not actually seek true North. Like any other compass, it seeks magnetic North. While the compass is prone to a host of errors, for most primary training in gliders it is not much used, and so will not be extensively discussed here. You will soon learn to navigate the local area by familiarity with its landmarks, and will generally be within gliding distance of your home airfield to boot. For the purposes of primary flight training in gliders, it is enough to note that the compass gives its most accurate reading when you are flying a straight glide at a constant airspeed.

The **variometer**, or sensitive rate-of-climb indicator, does not actually sense rate of climb or descent. (You expected this by now, didn't you?) Depending on its design, it measures airflow into and out of a fixed-capacity flask, or — in the case of certain electric variometers — senses the minute changes in electrical resistance caused by correspondingly minute changes in the size of a small blob on the circuit board. In either case, the vario indicates changes in ambient pressure. Varios tend to lag a bit in their response to pressure change — a consideration worth remembering when you are trying to center a small thermal.

The **yaw string** is simply a bit of string or yarn taped or tied somewhere on the nose or canopy of the glider in plain sight of the pilot. It indicates the direction of the airflow around it. It is the only flight instrument that has virtually no error or lag. The yaw string lets you know just how well you are coordinating stick and rudder as you fly the glider — a skill you will acquire in degrees as your experience increases.

Common errors

None

Completion standard

This lesson is complete when you can explain what each of the above instruments actually measures, and describe the sources of error for each one.

Glider assembly and disassembly

Lesson Glider assembly and disassembly

Purpose Gliders are occasionally disassembled for trailering or storage, then assembled for flight. Learning the manufacturer's recommended techniques for assembly (rigging) and disassembly (de-rigging) helps to avoid damage to the glider, saves time during rigging and de-rigging, and ensures correct control hookup.

Procedure Your instructor will guide you through the steps of de-rigging and rigging your training glider, including inspection of all critical parts such as spar pins, attach points, and control hookups.

Common errors Failure to procure and use the aircraft handbook

Failure to use the appropriate checklist during rigging or de-rigging

Failure to complete a positive control check after rigging

Completion standard This lesson is complete when you can:

- Describe the manufacturer's recommended sequence for de-rigging and rigging your training glider
- Inspect all attach points, assembly pins, and flight control hookups for conformance with the manufacturer's recommendations
- Perform a positive control check on the flight controls

The basics of free flight

Section 2

The free-flight portion of any glider flight is the portion of the flight between release from tow and entry into the airport traffic pattern preparatory to landing. You will probably be surprised, not to say delighted, to learn that your glider is quite stable and easy to fly safely — in fact, it is pretty much capable of flying itself for short periods of time. You also will learn that flying the glider with genuine precision is sometimes rather challenging.

The lessons in this section are varied. Some introduce you to aircraft systems, such as the primary and secondary flight controls. Other lessons concentrate on teaching flight maneuvers, such as turns and stalls. Finally, several of the lessons help you develop awareness of the "big picture" — some require you to recognize and deal with flight variables such as rising and sinking air or headwinds and tailwinds; some require keeping track of your location over the earth; and some require making sure that you remain within gliding distance of your home airfield.

As you progress through the lessons in Section 2, your instructor will gradually hand over more and more decision-making responsibility to you. Your goal is to acquire the necessary skills and judgment so that you can eventually take responsibility for all major decisions during the flight.

Primary flight controls

Lesson Introduction to the three primary flight controls: elevator for pitch control, ailerons for roll control, and rudder for yaw control

Purpose To demonstrate operation of the primary flight controls and their effect on aircraft attitude

Procedure Before flight, your instructor will explain the effect of elevator, ailerons, and rudder on aircraft flight attitude. In flight, your instructor will demonstrate the effect of the elevator by easing the stick forward to lower the nose. The glider will accelerate. Then your instructor will ease the stick back, and the glider will slow. When it is your turn, use light pressure on the stick first to lower, then raise the nose.

Ailerons are used to roll the glider wings into a bank so that a turn can be made. Your instructor will use left stick pressure to roll into a left bank, right pressure to roll into a right bank. After the demonstration, you will roll the glider into left and right banks.

Rudder is used to yaw the glider. Your instructor will demonstrate yaw to the left by applying left rudder pressure, and right yaw with right rudder pressure. Then you will do the same.

Common errors None

Completion standard You have completed this lesson when you can explain and demonstrate the effect of the three primary flight controls — elevator, ailerons, and rudder.

Other flight controls

Lesson Introduction to the other flight controls — elevator trim, airbrakes, flaps

Purpose To introduce the remaining flight controls and their use

Procedure After release, your instructor will demonstrate the use of elevator trim and its effect on pitch attitude and stick feel. You will learn to set a pitch attitude, wait a moment for the glider to stabilize, then trim off stick pressure using elevator trim. Your instructor also will demonstrate the operation of the airbrakes and familiarize you with their effect on the glidepath and airspeed of the glider, after which you will open and close them several times while flying the glider and noting their effect and feel. Finally, if your training glider is flap-equipped, your instructor will demonstrate flap operation, describing the flaps' effect on flight attitude, glidepath, and airspeed. Then, you will operate flaps and note their effects on the glider and on glider feel.

Common errors Forgetting to re-set trim after any appreciable pitch attitude change is made

Failure to look at each wing after operating flaps or airbrakes to confirm operation of the selected flight control

Failure to lock airbrakes closed or lock flaps up when finished operating them

Completion standard This lesson is complete when you can describe and demonstrate the operation of elevator trim, airbrakes, and flaps (if your glider is flap-equipped).

Scanning for traffic

Lesson Scanning for other air traffic

Purpose Collision avoidance is the most important job of every pilot. The best way to see (and avoid) other air traffic is to learn to look for it. This lesson teaches you what to look for and how to look for it.

Procedure Your instructor will show you the technique for scanning the sky around you. Select a sector of the sky about 30 degrees wide. Look carefully at it, paying particular attention to the portions just above the horizon, on the horizon, and below the horizon. Aircraft near the horizon are at an altitude similar to yours. As your eyes search the sector, fix your gaze near the horizon for a moment so that your peripheral vision will detect the motion of any object in the sector. Then select another sector and repeat the procedure. The goal is a frequent scan of all sectors visible from the glider cockpit.

If you detect a speck in motion in a sector, look closely and see if the speck has any apparent motion across, up, or down your canopy. You are most at risk from other traffic when the other traffic has no apparent motion on your canopy. The lack of apparent motion means that you and the traffic are converging. Alter your flight path as necessary in order to avoid a conflict.

Common errors Keeping the eyes moving in perpetual motion. This denies your motion-sensitive peripheral vision the opportunity to pick up the apparent motion of other traffic when it is still far off.

Completion standard This lesson is complete when you have acquired the habit of scanning all sectors for other air traffic.

Shallow banked turns and short straight glides

Lesson

Shallow banked turns left and right
Straight short glides

Purpose

Shallow banked turns and short straight glides introduce you to coordinated use of the primary flight controls: elevator, ailerons, and rudder. In addition, you will learn to precede each change in flight attitude with a careful visual check for other traffic.

Procedure

After release from tow, your instructor will demonstrate adverse yaw by using ailerons to bank the glider without applying simultaneous rudder. The nose of the glider will yaw in the opposite direction from the intended turn at first. Your instructor will then demonstrate use of the rudder in coordination with the ailerons to counteract adverse yaw. After seeing this demonstration it will be your turn to demonstrate adverse yaw, then demonstrate simultaneous use of rudder with aileron to counteract adverse yaw as you roll the glider into a shallow banked turn.

To be safe, you must look for other traffic before you begin any turn in a glider. In this lesson you will be taught to look carefully for other traffic before changing flight attitude. You also will learn to roll into shallow banked turns, stabilize the glider in the turn, and roll out of the turn. You will make several straight glides as well. During this lesson the airspeed indicator will be covered up. You will learn to control airspeed by monitoring the pitch attitude of the glider.

(continued)

Common errors

Forgetting to look for other traffic before applying the bank

Flying the turn at too steep a bank angle

Forgetting that each deflection of the ailerons must be simultaneously accompanied by rudder use to counteract adverse yaw

Too tight a grip on the control stick (Just relax and fly the glider with a light touch.)

Fixating on the yaw string

Completion standard

First and foremost, you will acquire the habit of looking for other traffic before you turn. This lesson is complete when you understand adverse yaw, use the rudder to counteract it as you enter and exit shallow banked turns, and can fly short glides with wings level.

Medium banked turns

Lesson Medium banked turns

Purpose Turning flight is a cardinal element of gliding. Accurate turns are required for a safe landing pattern. Climbing in thermals demands precise control of the glider in circling flight.

Procedure After clearing for other traffic, you will practice medium banked turns both left and right. Your instructor will emphasize clearing each turn, coordinated use of stick and rudder, and control of glider pitch attitude. You will discover the need for slight additional back pressure on the stick while circling to maintain target pitch attitude.

Common errors

Failure to look before turning

Forgetting the need for slight back pressure during the turn

Banking too steeply

Not using the rudder with the ailerons when rolling out of the turn

Forgetting to reduce back pressure on the stick when rolling out of the turn

Completion standard You will show mastery of this lesson when you precede each turn with a look around for traffic, coordinate the flight controls when entering a medium banked turn, maintain target pitch attitude during the turn, and use coordinated controls to roll out of the turn while maintaining the target pitch attitude.

Area familiarization, staying upwind

Lesson Airport area familiarization, landmark recognition, and staying upwind of your home field

Purpose Most new students occasionally feel a bit anxious or disoriented while airborne if they encounter difficulty locating the home airfield. The source of this anxiety is easily removed with a little training in landmark recognition and airport area familiarization. Wind awareness also is introduced at this time.

Procedure After release from tow, your instructor will familiarize you with prominent local landmarks and the location of your home airport relative to them. You also will learn the effect of wind on the ground track of your glider, and begin to take responsibility for planning your maneuvers so that you remain upwind of the airport within easy gliding range.

Common errors Forgetting to locate prominent local landmarks before trying to determine your position or locate your home airport

Inattention to the effect of wind on ground track and drifting downwind of the field as a result

Completion standard This lesson is complete when you demonstrate that you are oriented at all times, and plan your flights so that you remain upwind of the airfield, within gliding distance.

Straight-ahead stalls

Lesson Straight-ahead stalls

Purpose Stalls, with sufficient altitude, are safe. At low altitude they are not safe. In order to avoid a stall at low altitude, you will learn to recognize all the warning signs of an approaching stall while flying at a safe altitude. You will learn to stall the wings of the glider and make a normal recovery from the stall. You also will learn to precede each stall maneuver with clearing turns to avoid conflict with any other traffic.

Procedure After release from tow, perform clearing turns, looking out and down for other traffic. When satisfied the area is clear, your instructor will take the controls and slowly bring the glider to a stalled attitude while pointing out the warning signs of the approaching stall. (The warning signs of an approaching stall include having the stick back; a nose-high pitch attitude; reduced wind noise; sluggish control response, especially in roll; decreasing indicated airspeed; and pre-stall buffet.) Your instructor will gently stall the glider and demonstrate the recovery — easing the stick forward to reduce angle of attack. Then the sequence of events will be repeated as you follow through on the controls and list the warning signs of the approaching stall. Eventually, without assistance you will clear the area, list the warning signs as you approach the stall, stall the glider, and recover to a normal flight attitude.

Common errors First and foremost, failure to clear for traffic before stalling the glider

Forgetting to list the warning signs of the approaching stall

Moving the stick too far forward when recovering from the stall

Completion standard This lesson is complete when you: remember to clear the area, list the warning signs while approaching the stall, stall the glider, and recover with minimum loss of altitude.

Steep banked turns

Lesson Steep banked turns

Purpose Steep turns are sometimes required for climbing in small-diameter thermals. They also are necessary for changing the heading of the glider quickly.

Procedure After release from tow, clear the area for traffic. Steep turns begin like other turns but require more back pressure when the bank steepens to maintain target pitch attitude. You will first see a demonstration of a steep turn, then practice several yourself. You will learn to maintain target pitch attitude and bank angle during a series of steep turns left and right. Unless otherwise directed, each turn will comprise approximately 360 degrees of heading change — one complete revolution.

Common errors

Failure to clear before each turn

Insufficient additional back pressure as the bank steepens to prevent the nose from dropping

Forgetting to adjust ailerons as necessary to maintain target bank angle

Forgetting to release back pressure as you roll the glider back to wings-level flight

Completion standard This lesson is complete when you clear for traffic before each turn, roll into steep turns, maintain good pitch and bank control, and roll out within 10 degrees of a pre-selected heading.

Slow flight

Lesson Slow flight (a.k.a. flight at minimum controllable airspeed or MCA)

Purpose Safe flying demands avoidance of stalls at low altitude. This is best accomplished by thorough practice in recognizing the warning signs of an approaching stall. Slow flight is the name given to extended periods of flight just above the stalling speed while at a safe altitude. Slow flight gives you lots of opportunity to learn the stall warning signs, so that you will always recognize them and never inadvertently stall the glider while at low altitude.

Procedure After release and clearing the area, your instructor will slow the glider down and trim for slow flight. Your instructor will emphasize the warning signs the glider provides when flying at a high angle of attack. When you feel ready, you will fly the glider at a speed just above the stall. You will have as much time as you need to master recognition of all the signs of approaching stall. Slow flight maneuvers will include straight glides and shallow banked turns.

Common errors Forgetting to use elevator trim

Excess back pressure resulting in an inadvertent stall

Insufficient back pressure, resulting in failure to elicit from the glider all of the warning signs of flight at a high angle of attack

Banking the glider too steeply when turning in slow flight (In slow flight turns, the emphasis is on flying slowly, rather than on achieving a rapid rate of turn.)

Completion standard You are proficient in slow flight when you: remember to clear the area, use flight controls properly (including elevator trim) to maintain flight at a high angle of attack, and can — without stalling — fly straight glides and shallow banked turns while eliciting pre-stall warning signs from the glider.

Stalls from coordinated turns

Lesson Stalls from coordinated turns

Purpose This lesson teaches you to stall the glider from coordinated turns and recover with minimum altitude loss. You also will learn the signs of an approaching stall from a coordinated turn.

Procedure After release, clear the area for traffic. Once you have cleared the area, your instructor will demonstrate the procedure for stalling the glider from a shallow banked coordinated turn. Note the stall warning signs. Recovery requires release of back pressure as in wings-level stalls. You will then practice stalls from shallow banked turns.

After clearing again, your instructor will demonstrate stalls from medium banked coordinated turns. Note any change in the stall warning signs and in stall speed. Recovery, as always in a stall, requires reducing the angle of attack by easing the stick forward.

Stalls from steeply banked coordinated turns are usually difficult or even impossible to perform in modern two-seat gliders. Applying back pressure on the stick in a steep bank tends more to tighten the turn than to raise the nose of the glider and may result in a steep spiral. This makes for high stick forces and reduces the likelihood of safely achieving the wing stall angle without producing large, possibly dangerous "G" forces.

Common errors

Forgetting to clear the area before commencing turn stalls

Forgetting to keep the glider coordinated in the turn before inducing the stall

Using the ailerons to try to level the wings as the stall occurs

Pulling out of the post-stall dive too quickly, inducing a secondary stall

Completion standard You have completed this lesson when, after clearing, you can demonstrate stalls from shallow and medium bank coordinated turns with minimum altitude loss during recovery.

Stalls with airbrakes

Lesson Stalls with airbrakes

Purpose Stalls with airbrakes open introduce you to the stall characteristics of the glider when it is in the landing configuration (airbrakes open). You will discover whether airbrakes affect the warning signs that precede the stall, and whether they affect the stall speed of the glider. You also will learn that recovering from this type of stall with minimum altitude loss requires immediate closing of the airbrakes.

Procedure After release and clearing turns, your instructor will demonstrate straight-ahead stalls with airbrakes half open. When the stall occurs, your instructor will close the airbrakes, at the same time lowering the nose to break the stall. Both the control stick and the airbrake handle are moved in the same direction — forward — during recovery. Your instructor will demonstrate recovery with minimum altitude loss. You will then practice wings-level stalls and stall recovery, closing the airbrakes as you move the stick forward to break the stall, recovering with minimum altitude loss. Next, your instructor will demonstrate stalls with airbrakes from coordinated turns. When it is your turn, move the airbrake handle forward to close the airbrakes as you ease the stick forward to break the stall. With airbrakes closed and the nose of the glider lowered, gently ease out of the resulting dive.

Common errors

Failure to clear the area before practicing stalls

Forgetting to close the airbrakes during stall recovery

Completion standard Completion of this lesson requires you to simultaneously close the airbrakes and lower the nose to break a stall. Stall recovery will be accomplished with minimum altitude loss.

Crabbing during cruising flight

Lesson Crabbing for crosswind correction during cruising flight

Purpose Glider heading and glider track are not necessarily the same thing, due to the influence of any crosswind component present. This lesson teaches you to compensate for crosswind drift by judging and applying the approximate crab angle required by the wind.

Procedure You will make several long crosswind glides on a windy day.

Your instructor will point out that careful note of track over the ground will prove that the glider is not going, relative to the ground, where it is pointed. Instead, you will see that the glider is tracking downwind of the selected heading. Your instructor will demonstrate that, in order to make good the selected ground track, the glider heading must be adjusted to be upwind of the point on the horizon toward which you wish to track. This difference between glider heading and glider track is called the crab angle. The amount of crab required varies directly with crosswind component, inversely with glider airspeed.

Estimating the amount of crab required is done by simple trial and error. To apply the crab, simply make a shallow coordinated turn to point the nose a bit upwind of the horizon target point. Then, return to a wings-level coordinated glide with a constant heading and check the target point location on the canopy at brief intervals. If you see the target point moving along the canopy toward the downwind wing, you have applied more crab than is necessary; take some off. If on the other hand the target point is sliding along the canopy toward the upwind wing, then the initial crab angle is insufficient; add some more. During a long glide, crab angle is frequently monitored in this way, and corrected when indicated. Frequent monitoring is necessary during a long glide because the amount of crab required changes as crosswind changes, and the crosswind component frequently changes as you descend due to a number of weather factors. In addition, lift and sink encountered during the glide call for substantial airspeed changes, which in turn affect the amount of crab angle required.

Common errors

Many students initially confuse the crab with the slip. (Unlike the slip, during a crab the glider is flown in a coordinated fashion: yaw string straight back. Remember that the crab is a coordinated flight maneuver, and that correction for crosswind is made by altering the heading of the glider, not by flying it in a slip.)

Completion standard

This lesson is complete when you can show your instructor that you recognize situations when crabbing is called for in straight glides, and employ crabs in such conditions by estimating the amount of crab required, applying it, and making corrections as necessary to maintain a straight ground track toward a selected point on the horizon.

Flying in lift

Lesson Flying in lift

Purpose An enthusiastic appreciation of the possibilities of motorless flight really begins with your first climb in lift. For most people, this occurs as a result of an encounter with a thermal updraft. In this lesson you will learn the rudiments of flying the glider efficiently when in a thermal so that you can extend your flight.

Procedure After release, when a thermal is encountered, your instructor will clear the turn and then demonstrate a thermal climb by circling within the confines of the thermal. This requires close control of both pitch and bank. If pitch attitude is allowed to wander, the resulting uneven airspeed will result in an irregular circle and an inefficient climb; the thermal may even be lost altogether. Similarly, if bank angle is not held constant, rate of turn will vary, resulting in an irregular circle.

Most training gliders climb well in a thermal when the glider is flown slightly faster than minimum sinking speed for level flight. The extra speed is necessary because the additional load factor in a banked turn effectively increases the speed at which minimum sink occurs. Do not fly so slowly that you feel the pre-stall buffet, because the drag penalty for flight in this regime is very high and inefficient climb will result. Best control of airspeed, as always, is obtained by noting the target pitch attitude, then maintaining it as best you can.

Efficient climb in a thermal sometimes demands shifting the center of your circle one way or another in order to confine your flight path to the strongest part of the thermal. Your instructor will point out several cues that thermals often provide to remind you to shift your circle toward the center of the thermal.

Common errors

Forgetting to clear before rolling into the turn

Fixating on the variometer dial, simultaneously ignoring both other traffic and pitch and bank control

Flying too fast

Forgetting to trim the elevator as needed

Completion standard

You have completed this lesson when you can clear for traffic, enter a thermal, and climb, shifting the center of your circle occasionally if the thermal requires it.

Selecting cruise airspeed

Lesson Selecting cruise airspeed for maximum gliding distance in lift, sink, headwinds, and tailwinds

Purpose Most glider flying is done in the presence of wind, convection, or both. Wind and convection affect glide distance. To obtain the flattest glide angle in the presence of wind or convection during cruising flight, glider pilots adjust cruising speed after analyzing wind and convection.

Procedure During cruise, tailwinds and lift are your allies. They help you glide farther over the ground. Cruising at minimum sink speed keeps you in the air the longest possible time. The longer you stay aloft, the more time you provide your allies to help you glide the greatest possible distance over the ground.

Conversely, during cruising flight headwinds and sink are your enemies. If you fly too slowly, you allow them additional time to retard your progress over the ground. Thus, headwinds or sink require cruise speeds faster than best L/D if you wish to glide as far as possible over the ground. A useful rule for determining near-optimum airspeed with a headwind is to add the headwind velocity to the best L/D airspeed of the glider.

Determining the cruise speed to fly in sinking air is a bit more complicated. Because sink is your enemy, you know you must fly faster than best L/D speed. Exactly how much faster to fly can be determined with precision only after analyzing the performance curve of the glider you fly. Your instructor can suggest airspeeds suited to your particular training glider in various conditions of sinking air. Note that cruise speed to maximize gliding distance while in strong sink is much faster than best L/D speed for calm conditions. Although strong sink calls for high cruising speed to achieve the flattest glide angle, remember to respect your glider's rough air speed and maneuvering speed as conditions warrant.

Common errors

Flying too slowly in sink is common. This seems attractive to the new student, largely because the already strong sink indicated on the vario seems to warn against increasing the sink rate temporarily by lowering the nose. Analysis shows that flying slowly in sink is exactly the wrong thing to do!

Flying too slowly when headwinds are present is another common error.

Completion standard

This lesson is complete when you can select the airspeed that will maximize your glide distance in the presence of lift, sink, headwinds, or tailwinds.

Takeoff and aerotow

Section 3

Glider pilots and water-skiers have something in common: both must be towed. Both water-skiers and glider pilots must acquire the knack of handling things while on tow. Many students find this a bit frustrating at first, but the day will come when you acquire the knack, and then you will wonder why it ever seemed difficult. It's like riding a bicycle — a piece of cake, once you have the hang of it. So stay with it: You'll get it.

The lessons in this section teach you to deal with normal and emergency procedures during takeoff and while on aerotow. As much as your skills and knowledge allow, think ahead — it is your responsibility to bring your flight to a safe conclusion no matter what might happen to the towplane or towrope. Whether simulated or genuine, a low altitude emergency while on aerotow does not allow lots of time to make important decisions. For this reason, aerotow emergency procedures get a good deal of attention and drill in these lessons.

Finally, a word of warning. Since the vast majority of glider training flights in America are launched by aerotow, this book concentrates on teaching aerotow. However, there are other methods of launching. Ground launch can be accomplished with a specialized winch, or with a car or truck. Self-launching can be done in a motorized glider approved for self-launching. If one day you get the opportunity to learn either of these launching methods, make sure that you receive thorough training in normal and emergency procedures appropriate to the type of launch.

The takeoff checklist

Lesson The takeoff checklist: AABCCCDE

Purpose The simplest and best way to prepare for takeoff is with a checklist, to assure everything is in order for flight.

Procedure You will learn the AABCCCDE checklist as follows:

- A for Altimeter set
- A for Airbrakes closed and locked
- B for Belts on and secured
- C for Canopy closed and latched
- C for Controls free and correct, set for takeoff
- C for Cable (towrope) connected to the towhook
- D for Direction of wind
- E for Emergency Plan

Common errors Failure to memorize the checklist

Closing the canopy but forgetting to lock it down

Neglecting to check the belts of the other passenger/pilot

Forgetting to check elevator trim control or flaps during the Controls portion of the checklist

Completion standard This lesson is complete when you have committed the AABCCCDE checklist to memory and can perform the checklist in the cockpit without prompting.

Introduction to flying the aerotow

Lesson Introduction to flying the aerotow, using elevator and ailerons, and releasing from aerotow

Purpose In this lesson you will begin to learn the art of staying in place behind the towplane, using the ailerons and elevator. You will also learn the procedure for release from aerotow.

Procedure Your instructor will demonstrate takeoff and fly the aerotow until approximately 1,000 feet of altitude has been reached. At that time, your instructor will tell you to take control of the stick and use ailerons and elevator as needed to maintain position behind the towplane. Your instructor will continue to operate the rudder pedals, while you operate the ailerons and elevator. Your job is to maintain position on tow above the towplane wake, and to maintain position laterally by keeping the wings of the glider parallel to the wings of the towplane. You need not look out at the glider wings to do this — just use the left and right cockpit sills. Focus on the towplane, and you will see the cockpit sills in your peripheral vision.

If you find yourself out of position on tow, don't rush to regain correct position. First, prevent further deterioration of your position; then, when things have stabilized, ease back into position. Your instructor will coach you as the tow progresses, will assist you on the controls if need arises, and will let you know when it is time to release. When your instructor tells you to, pull the release knob, confirm that the towrope has disconnected, clear to the right, and make a gentle climbing right turn away from the towplane and rope.

(continued)

Common errors

Forgetting to keep the wings of the glider parallel to the wings of the towplane

Hurrying to return to position, resulting in excursions back and forth through and around the wake

After pulling the release knob, forgetting to confirm that the rope has actually released before initiating a right turn

Completion standard

You have completed this lesson when you can fly the glider using elevator and aileron on tow from 1,000 feet to 3,000 feet while your instructor minds the rudder for you, and when you can confirm release from tow, clear, and make a gentle right-hand climbing turn.

Flying the aerotow with stick and rudder

Lesson Flying the aerotow with stick and rudder

Purpose In the previous lesson you learned to fly the aerotow with two-axis control (elevator for pitch control, ailerons for roll control) while your instructor handled the rudder chores. In this lesson you add rudder control to aileron and elevator control. You will learn the amount of rudder needed to counteract adverse yaw while on tow, and will develop good coordination on tow.

Procedure Your instructor will perform the takeoff and climb until the glider is at 1,000 feet. When you are ready, your instructor will give you control of the glider — stick and rudder. Your instructor will coach you on use of the flight controls. Note the large amount of rudder needed to counteract adverse yaw — characteristic of most two-seat gliders, with their long wings acting as effective levers to yaw the glider whenever ailerons are deflected.

As before, concentrate on keeping the wings of the glider parallel with the wings of the towplane. Coordinate use of aileron and rudder. If you deviate from the target tow position, first prevent further deviation by stabilizing the situation, then ease the glider back into good position.

Common errors Forgetting to use rudder (or using insufficient rudder) whenever ailerons are deflected, resulting in substantial adverse yaw and repeated excursions across the wake

Completion standard You have completed this lesson when you can fly the glider using coordinated stick and rudder from 1,000 to 3,000 feet, release from tow, confirm that the rope is disconnected, clear, and make a gentle climbing turn to the right.

Shifting through the wake

Lesson Shifting through the wake between high and low tow

Purpose A by-product of towplane propeller thrust is the turbulent air behind and below the towplane, which glider pilots call the wake. This lesson familiarizes you with the location of the wake and introduces low tow position (flight beneath the wake).

Procedure Your instructor will demonstrate the location of the wake by easing the glider from high tow position lower and lower until the glider contacts the wake. The increase in noise and turbulence marks the boundaries of the wake very clearly. Then your instructor will ease the glider still lower, until the noise and turbulence decrease suddenly; this is the low tow position, with the glider on tow in smooth air slightly beneath the wake. While on low tow, note that you see the towplane from a much lower angle than in high tow. Transition to high tow will be made by easing the glider up through the wake.

Whether flying on high tow, through the wake, or on low tow, simply use the flight controls as before — coordinated stick and rudder to control the glider. When moving from high to low or vice versa, make changes gently and give the glider time to respond to your control inputs.

Common errors Students often rush the transition from one tow position to another. Hurrying from high to low can cause towline slack to develop as the glider nose is suddenly lowered, especially in the case of low-drag, high-performance gliders, which accelerate very easily. Hurrying from low to high causes a rapid increase in the induced drag of the glider wings, causing the towplane to lose speed, and in extreme cases to stall.

Completion standard You have completed this lesson when you can penetrate the wake to alternate smoothly between low and high tow.

Boxing the wake

Lesson Boxing the wake

Purpose The wake box maneuver requires excellent control of the glider while on tow as well as an understanding of the forces acting on the glider while in various positions relative to the wake of the towplane.

Procedure Your instructor will demonstrate a wake box. Note the slight bank angle needed to move the glider out to the side of the box and, once out to the side, to keep it there. Use the lift of the glider wings to move out to the side of the box by applying a slight bank. Along with the bank, it may be necessary to hold slight rudder pressure away from the towplane to counteract the tendency of the towline to pull the glider's nose back in line toward the towplane. This effect is most noticeable when the glider towhook is a nose hook, and much less noticeable if the glider has a belly hook.

As your instructor boxes the wake, you will see that the wake box is performed at a leisurely pace. At each of the four corners of the box, your instructor will pause briefly before continuing. When finished, the glider will have returned to the high tow position from which the maneuver began.

(continued)

Procedure continued

When it is your turn, use the controls gently and give the glider time to respond. Take your time. Pause for a moment at each corner, then ease the glider from one corner to the next. To get a sense of how far out or down you need to move the glider relative to the wake, use the changing perspective of the towplane features. Your instructor will coach you as you box the wake.

Common errors

Students commonly hurry the box. Instead, take your time and perform the box smoothly. Excess rudder use is common when trying to move out to the side of the box. Remember that the lift of the wings will do the work easily if you apply a little bank.

Completion standard

This lesson is complete when you can box the wake smoothly, relying on bank angle — not rudder — to move the glider out to either side of the box and to hold it there as required.

Takeoff and aerotow

Lesson Takeoff and aerotow

Purpose In this lesson you learn to fly the glider from aerotow start (takeoff) to finish (release). Your instructor will select a day without crosswind. A wing-runner will assist you at the beginning of the takeoff roll.

Procedure In your training you have seen a number of takeoffs performed by your instructor. The takeoff roll requires that you control pitch attitude with elevator and that you steer the glider on the ground using the rudder. In the absence of crosswind, keep the wings level as you roll. When the glider has reached flying speed it will fly itself off, if you have maintained the target pitch attitude. Stabilize the glider a few feet off the ground. Since you are now airborne, use the flight controls in a coordinated fashion: stick and rudder together.

The towplane requires a higher speed to get airborne than the glider does. Therefore, even after the glider has lifted off, the towplane is still on the ground, still accelerating to reach flying speed. Since the towplane is accelerating, so is the glider. This means that, if you hold a constant pitch attitude, as the glider accelerates it will start to climb, because even with constant pitch attitude the increase in airspeed translates into increased lift — and a climb. Therefore, to avoid premature climb, you must lower the nose very gently as the glider accelerates in order to maintain an altitude of three to five feet over the runway. Once the towplane lifts off, maintain the normal high tow position relative to the wake, using the now-familiar landmarks on your towplane for reference. Keep your wings parallel to the wings of the towplane. Release from tow when you have reached the predetermined release altitude.

(continued)

Common errors

Ballooning after takeoff is quite common in students' first few takeoffs due to the acceleration the glider undergoes until the towplane lifts off. The cure is to gently press the stick forward as necessary during acceleration in order to maintain a constant altitude (not attitude).

Completion standard

To complete this lesson you will fly the glider from takeoff to release altitude. A wing-runner will assist you in the takeoff.

Section 3

Introduction to premature aerotow release

Lesson Introduction to premature aerotow release procedures

Purpose An emergency premature release from aerotow is rare. However, if release occurs at low altitude, you have a limited time available for response. For this reason, premature release from aerotow receives a lot of attention in a glider flight-training program. In this lesson you learn to divide each aerotow into four stages and to have a separate plan for each stage in the event of early release from tow. Later in your training you will have opportunities to practice several low releases under the supervision of your instructor.

Procedure Premature release from aerotow may result from towplane difficulties, glider difficulties, rope breaks, or inadvertent release from tow. Glider pilot response depends on just how early the release is. There are four categories of early release.

The first category is release while there is still sufficient runway ahead of you to complete a safe landing. This may happen because the glider was improperly connected, the rope broke, or something in the takeoff or tow convinced you to release. Causes might include difficulty controlling the glider, an unfinished takeoff checklist, a loose canopy, a traffic conflict, or indeed anything else that you think may jeopardize the flight. The only safe thing to do in these situations is to release immediately! Whatever the problem is, it will be easiest and safest to deal with it after terminating the tow. Release and control the glider. If airborne, land and use the wheel brake to minimize ground roll. In any case, avoid the towplane and maneuver the glider clear of other obstacles.

(continued)

Section 3

Procedure continued

The second category presents itself when the takeoff has proceeded to the point where the glider can no longer be landed straight ahead on the remaining runway, but is too low to manage a 180-degree turn and fly back to the runway. This stage of takeoff will require an off-field landing pretty much straight ahead. You should have landing sites selected for this emergency by an early stage of your training. Some damage to the glider is possible, depending on your training site. The important thing is to avoid damage to yourself! Your instructor will provide you with suggestions about where the glider might best be landed in this event. **Do not stall the glider!** Fly it as best you can down to the ground. Injury is least likely to result if you maintain flying speed to touchdown.

The third category of early release is when the glider has sufficient altitude to make a 180-degree turn and return to the airfield, landing downwind. Textbooks often quote 200 feet above ground level as the minimum altitude for this maneuver, but the actual minimum altitude may be higher if the runway is very short, if the towplane is not very powerful, if the glider is heavy, or if weather conditions conspire to create a less-than-impressive climb angle as you depart the field. Your instructor will provide you with guidelines to help you judge the minimum safe altitude and maximum tailwind component for returning to the field and landing downwind safely.

The fourth and final category of early release occurs when the glider has sufficient altitude to make a more-or-less normal downwind-base-final approach, landing into the wind. Pay particular attention in this situation to what the towplane is doing. If the aerotow ended because of towplane problems, the tow pilot may be making a beeline for the airport. Plan your approach accordingly.

Common errors

The most common error is failing to have a plan for premature release. Plan your reaction to an early release before you climb into the cockpit.

In a low release followed by a 180-degree turn back to the airfield, students often make the turn downwind rather than into the wind.

Students may forget to mentally mark each of the four stages as the tow proceeds. Inattention to your location and lack of an immediate plan of action in an emergency wastes precious seconds and further restricts your field of action.

Completion standard

This lesson is complete when you have a plan for any premature release, can explain your plans to your instructor, and mark the transitions through the four categories of premature release on every launch.

Slack rope on aerotow

Lesson Slack rope on aerotow

Purpose Towrope slack can be induced by several factors, including rapid pitch attitude change, turning inside the towplane's radius of turn, towplane power reduction, and others. A large slack loop may present a hazard to the glider by threatening to entangle it. Slack loops of any size require care in restoring normal towrope tension.

Procedure Your instructor will induce several mild slack-rope conditions while on tow. During each recovery, note that the glider is moved out to the side a bit, and the nose of the glider is yawed away from the towplane with rudder. The slight surge on the towline when slack is taken up is partly damped by yawing both the towplane and the glider. In this way abrupt shocks on the rope are avoided, to the benefit of the rope, the glider, the towplane, and all pilots involved.

Slack is easily induced by diving a bit on the towplane. Higher performance gliders can induce slack more easily, because they offer less resistance to acceleration. If recovery from an initial slack-rope condition involves a very strong surge on the towrope, the elasticity of the rope can work like a slingshot and accelerate the glider, causing another round of slack rope and recovery. Using yaw to damp out the surge is an effective tool to prevent this cycle occurring.

Sometimes a very large slack loop can appear. If you are uncertain about your ability to recover from a large slack loop using normal means, pull the release and avoid entanglement with the end of the rope. In this situation it is usually best to pull up and away from the rope.

Common errors

Forgetting to yaw the glider away from the towplane with the rudder before the slack rope becomes tight again

Forgetting to neutralize the rudder just before the towrope becomes taut

Completion standard

This lesson is complete when you can induce, then recover from, slack-rope situations, using rudder as necessary to yaw the glider away from the slack rope.

Premature release and abbreviated pattern

Lesson Premature release from aerotow followed by an abbreviated pattern consisting of downwind leg, base leg, final approach, and landing

Purpose You should be prepared for an emergency at any time during aerotow. In this lesson you will deal with a simulated emergency during aerotow, releasing at an altitude and location that provide sufficient time for an abbreviated pattern.

Procedure During an aerotow, and quite likely without more than a moment's notice to you, your instructor will release the glider from the towplane. Altitude on this first simulated emergency release will be sufficient to allow an abbreviated pattern. Since you have already learned to have an emergency plan for each of the four stages of aerotow, you will fly an abbreviated approach and land the glider.

Common errors Failure to have a plan for premature tow release at any moment during the tow

Delay in turning toward the downwind leg

Forgetting to perform a downwind checklist

Forgetting to compensate for crosswind

Insistence on flying a large, square pattern even if the situation requires an abbreviated pattern

Losing track of the location of the towplane (The towplane pilot might want to land quickly also, which would likely be the case in a real emergency.)

Completion standard This lesson is complete when you demonstrate that you have a plan for an emergency on tow at an altitude that allows an abbreviated pattern, and can carry it out without delay in a simulated emergency.

Low release and downwind landing

Lesson Low release from aerotow, 180-degree turn, and downwind landing

Purpose A low release from tow and subsequent downwind landing provide a realistic emergency simulation. Execution of the maneuver demands good decision-making and competent handling of the glider, all while at low altitude.

Procedure As in all takeoffs, note the direction of the wind during your takeoff checklist. Wind direction at takeoff determines not only the corrections you must make for crosswind or other conditions but also the direction you select to turn in the event of a low release and 180-degree turn to the airfield. It is best to turn into the wind during this maneuver to preclude being blown downwind of the field while making the turn.

Fly a normal takeoff and do your best to maintain the track directly over the extended imaginary centerline of the runway behind you as you climb out. A minimum altitude of 200 feet above ground level is required for this maneuver. Your instructor will release the glider from the towrope. When you see the rope snap away, make a coordinated turn into the wind using approximately 45 degrees of bank. **Maintain flying speed.** Line up on the runway, correct for any crosswind, unlock the airbrakes, and use them as necessary. The downwind landing will require a somewhat lower approach angle than you are accustomed to flying when landing into the wind. After the flare and touchdown, use the wheel brake both to shorten your landing roll and to avoid the ground loop that can follow a downwind landing as control authority decays.

(continued)

Common errors

Neglecting to decide before takeoff which way to turn back to the field in the event of a low release

Forgetting to call out "200 feet" (decision height for a 180-degree turn) as you pass through that altitude

Using too shallow an angle of bank for the 180-degree turn back to the airfield

Flying the final approach too high, resulting in a long landing because of increased ground speed caused by a tailwind

Completion standard

This lesson is complete when you can make all the decisions noted above and fly the glider to a safe downwind landing without assistance.

Crosswind takeoffs

Lesson Crosswind takeoffs

Purpose Mother Nature rarely provides a good soaring day without wind. Most days, the crosswind component is large enough to require counteracting response from the glider pilot.

Procedure A crosswind takeoff presents two problems. First is the effect of crosswind on the glider when it is in motion but has yet to lift off. Second is the effect of crosswind as you try to maintain the runway centerline after liftoff.

Start the takeoff roll with downwind rudder. If, for instance, the wind is blowing from left to right, start the roll with full right rudder. This minimizes the glider's tendency to turn, or weathervane, into the wind (to the left in this example). As speed and rudder authority increase, you can ease off downwind rudder pressure and steer the glider more or less normally with the rudder. Use ailerons to keep the upwind wing a bit low. When the glider is clearly ready to fly, fly it off and keep it off.

After takeoff, you no longer can count on the friction of the tire with the ground to keep the glider from drifting downwind. To prevent drift, keep your upwind wing slightly low as you did during the takeoff roll, and use rudder as necessary to maintain ground track. Once you have cleared the end of the runway and any obstacles near the sides of the field, level the wings as you neutralize rudder, and fly a normal coordinated tow to altitude.

Common errors Failure to note crosswind direction before takeoff

Failure to apply downwind rudder at the beginning of the takeoff roll

Forgetting to keep the upwind wing slightly low

Lifting the glider off too soon, resulting in settling back to the runway and a considerable side load being applied to the landing gear

Completion standard You have completed this lesson when you can plan for crosswind takeoffs and can maintain ground track over the centerline of the runway.

Takeoff without a wing-runner

Lesson Takeoff without a wing-runner

Purpose Occasionally during your gliding career you may lack the services of a wing-runner to help you launch. Most two-seat training gliders can be launched unassisted without too much difficulty. In this lesson you will learn the technique for launching without assistance. (Make a note, however, that not all gliders are suitable for this type of launch. In particular, single-seat racing gliders have poor aileron control at low speeds and are much better off being launched with the assistance of a wing-runner.)

Procedure Launching without a wing-runner requires that full aileron deflection be used from the start to raise the low wing off the ground. Because of the glider's tendency to yaw toward the wing that is touching the ground as the takeoff roll begins, full rudder toward the high wing also will be needed. In many gliders, the rudder will become effective before the ailerons do. As the rudder authority increases, steer the glider on the ground with the rudder and continue to hold full aileron until the low wing lifts off and the wings are nearly level. Once you have accomplished this, the rest of the takeoff roll and liftoff will be identical to an assisted takeoff under similar conditions.

Unassisted takeoffs in a crosswind also can be accomplished. Two schools of thought exist on what technique should be employed. The first teaches that the upwind wing should be lowered onto the ground before takeoff, and aileron and rudder should be used exactly as described above.

The second school teaches that the downwind wing should be lowered to the ground before launch. To support this, devotees note that the tendency of the glider to veer (or weathervane) into the wind during a crosswind launch is now counterbalanced by the yaw resulting from wingtip drag on the ground on the downwind side of the glider. The opposition of these two forces makes steering the glider while on the ground easier. The price for achieving this advantage, however, is that the crosswind now is resisting the effort of the pilot to lower the upwind wing. If the ailerons of the glider are sufficiently powerful to overcome this resistance, then the launch likely will continue without difficulty. However, if aileron control is poor, or if the crosswind is strong, it may be quite difficult to lower the upwind wing until considerable speed has been attained.

Consult with your instructor about the preferred technique for your particular training glider.

Common errors

Students tend to hold full rudder too long, causing the glider to steer off to the downwind side of the runway. In most gliders the rudder gains authority before the ailerons do. As a result it is usually necessary to reduce rudder deflection while still holding full aileron to raise the low wing.

Students often hold full aileron too long, causing the upwind wing to rise and continue to rise until the opposite wingtip contacts the ground.

Completion standard

You have completed this lesson when you can launch the glider without a wing-runner, using the control techniques described above.

Pattern and landing

Section 4

It is axiomatic in flying that a good traffic pattern makes for a good landing. This is particularly true of gliding, since in a non-motorized glider you must do the approach and landing correctly the first time. This sounds daunting at first, but you will learn that it really is not terribly complicated or even particularly difficult. All that is required is an observant and self-disciplined pilot and a glider that has adequate glidepath control. (Your training glider should provide sufficient glidepath control by means of spoilers or airbrakes.)

Over the years I have flown thousands of patterns and landings in gliders. As an instructor I have witnessed many thousands more. From all this I have concluded that the key to flying a good pattern and landing in a glider is having the self-discipline to enter the traffic pattern when good sense tells you it is time to do so. If you delay pattern planning and entry until it is too late, you will find yourself with a rapidly diminishing list of safe options. You will cease feeling like a pilot-in-command and will begin to resent having placed yourself in a perilous situation unnecessarily. So, make it your policy to think ahead and keep track of your options. Make the decision to head for the airport when the situation requires you to do so. Accept the fact that concluding your flight safely is preferable to taking excessive risks at low altitude in an attempt to lengthen or "save" it. After all, the cost of another launch is a bargain when you compare it to the cost of making repairs to the glider — or to you.

Finally, remember that the traffic pattern is a tool, not an end in itself. It is the tool we use to set up a safe landing. Act promptly to adjust or modify any traffic pattern if doing so will make a safe landing easier.

The landing checklist

Lesson The Landing checklist: FUSTALL

Purpose The simplest and best way to prepare the glider for landing is with a memorized checklist. The FUSTALL landing checklist is adequate for almost any glider not equipped with a motor.

Procedure You will learn the FUSTALL checklist:

- F for Flaps set for landing
- U for Undercarriage down and locked
- S for Speed appropriate for approach
- T for Trim set
- A for Airbrakes checked for proper operation
- L for Look for other traffic
- L for Land the glider

Note that even if your training glider does not have flaps or retractable landing gear, do not eliminate those items from the checklist. If you repeat them even when you don't need them, you will be more likely to remember them when you do. Practice the checklist until you can do it in 10 or 12 seconds. Taking any longer than this usually means you are neglecting the critical tasks of looking for other traffic and judging your approach to land. It's best to do the checklist immediately after entering the downwind leg of the traffic pattern. When checking flaps or airbrakes, make sure that you have moved the control you intended to move by observing the wing.

Common errors

Failure to memorize the checklist

Taking too long to execute the checklist

Doing the checklist too soon (five miles out) or too late (on short final approach)

Forgetting to lock the landing gear after lowering it

Failure to look out at the wing as you operate the flap or airbrake handles

Completion standard

This lesson is complete when you have committed the FUSTALL landing checklist to memory and can perform the checklist without prompting at the appropriate point in the landing pattern.

Introduction to the landing pattern

Lesson Introduction to the landing pattern

Purpose Landing a glider accurately requires that you know and understand the procedures employed to make each landing a success. You also must possess the discipline to employ them in every approach you fly. This lesson introduces the principal decisions and procedures used to assure a safe pattern and landing.

Procedure Your instructor will discuss, then demonstrate, the components of a normal pattern and landing. These include: assessment of local weather conditions, selection of an approach speed, arrival at the initial point, flying the entry leg, entering the downwind leg, performing the landing FUSTALL checklist, flying the base leg, flying final approach, flaring, touchdown, rolling out, and securing the glider out of harm's way.

Common errors None

Completion standard This lesson is complete when you can describe accurately each event listed in the procedure section above.

Normal pattern and landing

Lesson Normal pattern and landing

Purpose Gradual transfer of responsibility for the pattern and landing from the instructor to you

Procedure You will fly many traffic patterns and landings in this stage of your training. Your job is to learn the sequence of decisions necessary for a safe pattern, then gradually take responsibility for making those decisions without assistance or prompting. To simplify this task a bit, your instructor will operate the airbrakes in your first patterns, leaving you free to concentrate on pattern planning, the landing checklist, angle judgment, pitch control, and clearing for other traffic. As you acquire skill and experience in pattern planning and landing, your instructor will encourage gradual transfer of all responsibility for a safe pattern and landing to you.

Patterns and landings flown as part of this lesson will include patterns with the airspeed indicator and altimeter covered, and patterns with combinations of light to moderate winds and convective conditions.

(continued)

Common errors

Failure to memorize and practice the FUSTALL landing checklist

Performing the landing checklist someplace other than on the downwind leg

Fixation on executing the pattern, to the exclusion of any awareness of other traffic

Excess reliance on the altimeter (Concentrate instead on keeping your intended touchdown spot within easy gliding distance.)

Forgetting to establish any crab required by wind conditions

Taking your left hand off the airbrake handle once the airbrakes have been unlocked and checked in the FUSTALL checklist

Gradually adding back pressure on the stick during the base leg and final approach because you are a bit anxious (a wrong-headed tendency that must not be allowed to develop into a habit)

Completion standard

This lesson is comprised of a series of patterns. You have completed this lesson when you can, without prompting, fly normal patterns and landings, taking complete responsibility for planning and executing the pattern and landing and securing the glider.

Introduction to the forward slip

Lesson Introduction to the forward slip

Purpose The forward slip can be used during a landing approach to dissipate excess altitude without increasing speed. The cross-control flight attitude presents a substantial portion of the fuselage side to the relative wind, causing an increase in parasite drag and a steeper descent than would otherwise be the case.

Procedure At altitude, after clearing, your instructor will demonstrate the forward slip by gently applying left rudder pressure simultaneously with right stick pressure while maintaining a constant pitch attitude. The glider will then be in a bank to the right, but no turn to the right will follow because the rudder is resisting the turning force of the wings. Note the yaw string, which will be swept toward the left side of the glider due to the sideways path of the aircraft through the air. Since the glider is going somewhat sideways, take a moment to note that the glider is not going where it is pointed. Check the variometer to see if the rate of sink has increased from level flight, and if so by how much. Check also to see if the indicated airspeed is the same as it was in level flight.

After returning to wings-level flight your instructor will demonstrate a forward slip to the left — that is, left bank will be applied simultaneously with right rudder. After returning to normal flight you will try the forward slip yourself, in both directions, with and without airbrakes.

(continued)

Procedure continued

Note that the effectiveness of the forward slip varies widely from one model of glider to another. Some gliders exhibit a large increase in sink rate, while others are only minimally affected. Learn the forward slip characteristics of any glider you fly in the future by practicing the maneuver at altitude. It is too late to find out during a too-high final approach that your glider does not slip well! Note also the effect of the forward slip on indicated airspeed. Some gliders have pitot systems that introduce large indicated airspeed errors during a slip.

Common errors

Students often forget to maintain a constant pitch attitude. Remember to keep the nose down somewhat, because the forward slip is used to dissipate altitude most often when on approach, at which time a stall is dangerous.

Students may apply more bank than the rudder can "hold off." In most gliders, rudder authority is limited, and if too steep a bank angle is used the turning power of the wings will defeat the resistance of the rudder and cause the glider to turn slowly in the direction of the low wing.

Completion standard

This lesson is complete when you can describe and demonstrate a forward slip at altitude, with and without airbrakes, in either direction (left wing low, right rudder; right wing low, left rudder).

Forward slip in the pattern

Lesson Using the forward slip for glidepath control in the traffic pattern

Purpose The forward slip increases the drag of the glider. It can be used in the traffic pattern to steepen the approach path.

Procedure The forward slip will be used on base leg and final approach as needed to steepen the glidepath of the glider. Use the slip as much as possible as your primary glidepath control during this approach. Because different kinds of gliders vary considerably in the effectiveness of the forward slip, the decision whether to use airbrakes to supplement the slip for this lesson will be made by you and your instructor. Maintaining the target pitch attitude as you apply and hold the slip is very important. Excess airspeed will result in consuming lots of runway during the flare. Adjust the amount of slip as necessary to maintain the glidepath you want. As you approach the ground, be prepared to take off the slip to avoid landing the glider with a substantial side load on the landing gear. If you are not using airbrakes during the approach, anticipate a longer than usual float in ground effect. Do not force the glider to land at too high a speed, but rather hold it off and let the speed bleed away. After landing, open the airbrakes fully and use the wheel brake as necessary.

Common errors Failure to maintain the runway centerline while on final approach

Indecision as to which wing to lower when applying the slip (decide this before you enter the pattern)

Waiting until too close to touchdown before beginning to remove the slip, which may impose heavy loads on the landing gear

Completion standard This lesson is complete when you can apply a forward slip to increase the descent rate of the glider in the pattern and land the glider after returning to coordinated flight.

Introduction to the sideslip

Lesson Introduction to the sideslip at altitude

Purpose The sideslip can be employed to prevent drifting downwind when landing in a crosswind.

Aerodynamically, the sideslip is identical to the forward slip. In application, however, the presence of a crosswind component makes it possible for you to slip in such a way that the glider's ground track matches its heading. The glider pilot lowers the glider's upwind wing and applies downwind rudder so that the tendency of the glider to slip sideways exactly matches the crosswind component. This allows the practiced pilot flying the approach to line up on the extended centerline of the runway and maintain ground track over that extended line all the way to the flare point.

Procedure On a breezy day, at altitude, your instructor will fly a straight glide such that the heading of the glider is at a right angle to the wind at your altitude. Note the drift of your glider over the ground: The crosswind causes your ground track to diverge downwind from the heading of the glider. By gradually lowering the upwind wing and applying downwind rudder, your instructor will demonstrate a sideslip that will exactly counteract the drift imposed by the wind, and you will be able to see that your heading and your ground track are now the same. After returning to a straight glide, you will take the controls and try it yourself. After a few tries, reverse the heading of the glider so that the wind aloft is now coming from your other wing and try a few sideslips in this condition.

Common errors Forgetting to maintain a lookout for other traffic

Using more bank than the rudder can hold off, resulting in a slow turn to windward

Allowing the nose to rise during the slip, resulting in decreased airspeed

Completion standard You have completed this lesson when you can apply just enough sideslip to erase any difference between heading and track in a crosswind.

Crosswind patterns and landings

Lesson Crosswind patterns and landings

Purpose Weather conditions seldom arrange themselves so that the wind is right down the runway. Most landings must be made with a crosswind component. This lesson presents the pattern adjustments and flying techniques that make planning and flying crosswind approaches and landings pleasant and safe.

Procedure Crosswinds oblige the pilot to make substantial adjustments to the traffic pattern. These adjustments include altering the spacing of the downwind leg relative to the field, crabbing into the wind to maintain a parallel ground track during the downwind leg, altering the timing of the turn to base leg, crabbing into the wind on base leg, altering the timing of the turn to final approach, crabbing (or sideslipping) on final approach, and adjusting the degree of crab or sideslip during final approach to compensate for wind gradient effects. Landing the glider should be accomplished without side loads on the landing gear, and the glider needs careful control during the rollout to prevent a ground loop.

Your instructor will require you to plan your patterns with these adjustments in mind. Before you enter the traffic pattern, observe local wind indicators closely and anticipate the crab angles and pattern spacing adjustments that conditions demand. You will probably be surprised to learn just how large the adjustments are that glider pilots must make to allow for crosswinds in the landing pattern. Your instructor will coach you as you plan, and then fly, crosswind patterns and landings.

(continued)

Common errors

Failure to plan for the effect of the wind before entering the pattern

Forgetting to use a crab on the downwind leg and drifting either too close to, or too far from, the field as a result

Applying a crab angle that is too small for conditions, particularly on the downwind and base legs

When using a sideslip on final approach, lowering the downwind wing rather than the upwind wing

Completion standard

This lesson is complete when you can plan for the effects of crosswind on your pattern and landing, adjust your pattern spacing accordingly, and employ the crab and the slip as needed to complete the pattern and landing.

Accuracy landing

Lesson Accuracy landing with rollout to a pylon

Purpose Accuracy landings demand close management of the energy of the glider in the pattern. By training to a high standard of skill in the pattern and landing, you provide yourself with the widest possible margin of safety for solo flights at your home field. You also prepare yourself for the future demands of badge flights, cross-country flights, landings at unfamiliar airports, and off-field landings, as well as for the accuracy landing requirements of your Private Pilot checkride.

Procedure Before launching, you and your instructor will place a marker on the runway or rollout area. During the subsequent approach and landing, your job is to bring the glider to a stop short of, but within 100 feet of, the marker. You may not use excessive force on the controls, force the glider to land prematurely, or abuse the wheel brake. Rather, you must plan and execute the approach so that only moderate control pressures and reasonable braking are necessary to bring the glider to its appointed spot.

Common errors Failure to correct problems or miscalculations early even when you detect them early enough to undo them with minimum effort

Trying to force the glider to land rather than flaring properly when it becomes clear that the glider has too much energy and may pass the pylon

Applying too much force to the wheel brake

Completion standard This lesson is complete when you can consistently roll up to within 100 feet of the pylon without requiring hard or violent use of the controls.

Landing over an obstacle

Lesson Landing over an obstacle

Purpose A general rule of thumb in landing over an obstacle is that the obstacle prevents you from using the runway beyond it for a distance 10 times as long as the obstacle is high. Because you are left with less runway than you otherwise would have, landing over an obstacle provides you an opportunity to show all you know about energy management in the pattern. You must plan your pattern to take into account wind and convection and the height of the obstacle. In flying your pattern, you must use precise airspeed control and a forward slip with airbrakes (once you are sure you have cleared the obstacle), then flare, touch down, and roll out using minimum runway.

Practicing landing over an obstacle prepares you not only for patterns at airports with nearby obstructions but also for off-field landings where obstacles such as trees, power lines, or buildings may be present.

Procedure After a thorough pre-flight briefing, your instructor will fly with you and demonstrate landing over an obstacle. (Whether the obstacle is simulated or real is up to your instructor and your airport environment.) Note the importance of assessing wind and convection and flying your pattern accordingly. Your instructor will make sure the obstacle has been cleared, use airbrakes and a slip as necessary to descend steeply, and maintain sufficient airspeed so that a short flare can be made, avoiding a hard landing.

Next it will be your turn. Energy management in the pattern is very important, because insufficient energy may not allow you to clear the obstacle, while excess energy may be difficult to dissipate before you use all remaining runway. Your number one job in an obstacle approach is to clear the obstacle! Once you know that you can do that, employ airbrakes and a slip as necessary to steepen the descent path over the obstacle. Maintain sufficient airspeed to allow for a brief flare, or a hard landing may result. Once the wheel touches down, open the airbrakes fully and use the wheel brake as your instructor advises.

Common errors

Imprecise pitch attitude/airspeed control

Indecision as to which wing to lower in the slip

Forgetting to use full airbrakes before applying the slip

Flaring abruptly, causing ballooning and the use of additional runway

Completion standard

This lesson is complete when you can demonstrate consistent planning and good control of the glider when landing over an obstacle, using airbrakes and a slip as needed, and using the wheel brake to minimize rollout.

Section 4

Advanced free-flight maneuvers

Section 5

The maneuvers in this section are a natural extension of the flying skills you learned in Section 2 of this book. The attention given to precision turns and zigzag turns helps you improve your coordination of the controls and your control of the glider. One of the great joys of flying gliders is developing and maintaining genuine precision in making the glider do what you want, when you want.

Incipient spins, spiral dives, and unusual attitude recoveries will present you with greater excursions in pitch attitude, bank angle, and load factor ("G" force) than you have likely experienced before. Your instructor will introduce these maneuvers in your training program at the appropriate time. Do not practice them in solo flight unless you have received express permission to do so from your flight instructor and have an endorsement to that effect in your pilot logbook.

Precision turns

Lesson Precision turns

Purpose Precision turns develop your ability to maneuver the glider with skill and grace. Satisfactory performance of these turns requires you to think a bit ahead in order to make the glider fly as you wish it to fly. This ability to anticipate and then make control inputs is central to acquiring good control of the glider. Your instructor will demonstrate several medium bank 360-degree turns, clearing before each turn. Note that each turn begins and ends with the glider pointed at the same landmark on the horizon, and that pitch attitude and airspeed are constant during the turn. Note also that the rollout to wings-level flight must begin well in advance of the glider having completed 360 degrees of turn, or the glider will overshoot the target landmark by a considerable amount. The amount of anticipation required varies directly with the angle of bank employed.

Since you already have mastered shallow, medium, and steep bank turns in a previous section of your training, you will be free to concentrate on the precision with which you roll out of the turn on a pre-selected heading.

Procedure Your instructor will request various combinations of bank angle (shallow, medium, steep) and heading change (360 degrees, 180 degrees, or simply rolling out and pointing at a selected prominent landmark).

Common errors

Forgetting to clear the turn before applying the bank

Monitoring the airspeed indicator too closely, rather than monitoring pitch attitude

In the case of steeply banked precision turns, entering the steep bank with insufficient airspeed to guarantee good controllability of the glider during the turn

Failure to initiate rolling out of the turn until the target landmark has already appeared straight ahead

Completion standard

This lesson is complete when you demonstrate ability to roll out within 10 degrees of a pre-selected landmark on the horizon from a shallow, medium, or steep turn. During each turn you must also demonstrate good pitch and bank control.

Spiral dive entry and recovery

Lesson Spiral dive entry and recovery

Purpose In this lesson you learn the distinguishing characteristics of the spiral dive, how to enter and recover from one, and the hazards associated with allowing a spiral dive to go unchecked.

Procedure Your instructor will clear the area, then begin a spiral dive. The spiral dive is exactly what the name implies — a coordinated turn combined with rapid loss of altitude. The bank employed is quite steep — about 70 degrees. Naturally, a bank this steep exerts a considerable load factor ("G" force) on the glider and its occupants, which in turn substantially increases the glider's rate of sink — hence the diving flight path with rapid altitude loss. During the spiral dive, your instructor will draw your attention to several characteristics of the spiral dive. They include the high indicated airspeed (well above stall speed), high "G" forces, relatively loud wind noise, and crisp control response, consistent with the high speed of the glider. In addition, during the spiral dive you will learn that the effect of backward motion on the control stick will not raise the nose of the glider (and consequently is useless in trying to recover from the dive). This is because the steep bank means that any increase in angle of attack of the wing will generate more lifting force inward — tightening the turn — than it will upward. Tightening the turn only results in greater loads, airspeed, and sink rate, and eventually could lead to failure of the glider.

Recovery from the spiral dive is simplicity itself. End the spiral, or turn, by rolling out of the bank. Then gently ease out of the dive.

Common errors

Forgetting to clear before beginning the spiral dive

Failure to apply sufficient angle of bank to enter the spiral

Applying excessive force to stick and rudder to roll out of the bank (Remember that the spiral dive is a high-speed maneuver, and that speeds in excess of maneuvering speed may well be encountered, resulting in high loads on the glider.)

Using excessive back pressure on the stick when pulling out of the dive, inducing high loads on the glider and possibly causing a high-speed stall

Completion standard

This lesson is complete when you can describe the spiral dive with its characteristics and associated hazards, clear the area, and demonstrate entry and recovery without inducing excessive loads on the glider.

Incipient spin entry and recovery

Lesson Incipient spin entry and recovery

Purpose The spin is a variety of stall. As with any stall maneuver, there are two principal reasons for studying and practicing it. The first is to learn the warning signs that the glider provides before the stall occurs so as to prevent inadvertent stalls, particularly at low altitude. The second is to learn the recovery procedure that returns the glider to normal flight with minimum altitude loss.

Procedure After clearing, your instructor will demonstrate a spin entry by flying the glider slowly, in a shallow bank, while using excess rudder in the direction of the turn. Skidding a turn results in the wings creating unequal amounts of lift. When the stall angle is reached, simultaneous pitching, rolling, and yawing motions combine to cause the glider to rotate with a nose-down attitude, with a very high sink rate. Your instructor will point out to you that load factor and airspeed are both rather low during the spin. Recovery will be made in accordance with the recommendations in the flight manual for your particular training glider. Generally, these consist of first using full opposite rudder to stop the rotation, then, with ailerons neutral, easing the stick forward to break the stall while neutralizing the rudder. With the wings unstalled, the glider can be eased out of the resulting dive by applying gentle back pressure on the stick.

It is important to remember that the spin is a variety of stall. Because the wing is stalled, it is impossible for the glider to develop much airspeed during the spin due to the substantial wing-induced drag. Consequently, the airspeed during the spin will be below maneuvering speed in almost all gliders, and load factor also will be low. It is only during a too-vigorous pullout from the resulting dive that high loads are induced.

Common errors

Failure to study your glider manual's spin-recovery recommendations

Forgetting to clear the area before entering the spin

Reluctance to raise the nose high enough, and skid the turn forcefully enough, to induce a spin entry

Attempting during the spin to raise the low wing with the ailerons (Remember to keep the ailerons neutral until after the stall is broken.)

Completion standard

You have completed this lesson when you can explain the cause of the spin, explain entry and recovery procedures, and demonstrate entry and recovery without hazard to your glider, its occupants, or other aircraft.

Zigzag turns

Lesson Zigzag turns (alternating left and right turns of 45 to 90 degrees of heading change as your instructor requests)

Purpose Gliders, with their long wings, exhibit considerable adverse yaw when ailerons are deflected. Applying just enough rudder to counteract adverse yaw makes for clean, low-drag flying. Zigging and zagging require you to roll into and out of banked turns every few seconds. Doing so gives you plenty of opportunity to acquire a feel for how much rudder is needed to counteract adverse yaw. You also will employ skills already firmly established, such as clearing each turn and rolling out of each turn on a precise heading or landmark.

Procedure After clearing, your instructor will demonstrate several zigzag turns. Notice that the turns are entered crisply, and that only a small amount of heading change occurs before the glider is banked (after clearing) to turn in the opposite direction. The turns are frequent because the object of this flight lesson is to perfect your coordination skills rolling into and out of banks. When your instructor gives you the controls, use medium bank turns and select and maintain an airspeed a bit faster than the speed for best L/D in calm air. Eventually, you can practice zigzag turns with shallow banks, then with steep banks. Note that the amount of rudder required changes with the airspeed at which any zigzag is performed. As a general rule, adverse yaw is greatest when the wing is flown at a high angle of attack (low speed), and diminishes at higher speeds. Maintaining the target pitch attitude and airspeed during each series of turns simplifies your task considerably.

Common errors

Forgetting to clear the area before beginning

Forgetting to clear each turn in the series

Failing to apply rudder pressure simultaneously with aileron pressure

Forgetting to maintain constant pitch attitude during each series of turns

Completion standard

To complete this lesson you must remember to clear all maneuvers, demonstrate good control of pitch attitude, and enter a series of alternating left and right precision turns while showing good coordination and heading control.

Section 5

Unusual attitude recovery

Lesson Unusual attitude recovery

Purpose In this lesson you demonstrate your ability to recognize and recover from unusual flight attitudes. After establishing an unusual attitude, your instructor will turn control of the glider over to you. You must decide what attitude the glider is in, then recover to a wings-level glide. Because your instructor wants to know that you can handle any of these situations without assistance, your instructor will not identify the attitude the glider is in when control is handed over to you. That will be your responsibility. Just apply your knowledge to diagnose each attitude, then apply your experience to recover promptly with minimum altitude loss.

Procedure At altitude, and after releasing and clearing the area, your instructor will tell you that the glider will now be placed in an unusual attitude. Your instructor will not identify the attitude by name. Once the attitude is established, the glider will be turned over to you with the words "You have the glider." You will then recover from the unusual attitude, restoring the glider to a wings-level glide at moderate airspeed (best L/D speed will be fine). When you have recovered, your instructor will take control of the glider with the words "I have the glider," clear the area, and put the glider into another unusual attitude. As before, you will be given control of the glider with the words "You have the glider." You will then recover to a wings-level glide. Before the lesson is over, your instructor will have presented you with a variety of unusual attitudes such as stalls, dives, climbs, steep turns, spins, and spiral dives. In each case, your job is to recognize the attitude the glider is in, then smoothly recover with minimum altitude loss.

Common errors

Forgetting that altitude is being used quickly (Make a point to keep oriented and to stay within gliding range of your home field.)

Forgetting to keep a careful lookout for other traffic

Completion standard

You have completed this lesson when you can identify and recover from unusual attitudes with minimum altitude loss and without prompting or assistance from your instructor.

Section 5

Appendix

Minimum requirements for solo flight

These guidelines are for reference only. Consult current Federal Aviation Regulations before solo.

Aeronautical knowledge

The student must demonstrate to an authorized instructor satisfactory knowledge of the portions of FAR parts 61 and 91 that are applicable to student pilots. This demonstration must include administration of a written exam, to be graded by the instructor authorizing solo flight. The exam must include questions on applicable regulations and on the flight characteristics and operational limitations of the glider to be flown solo.

Pre-solo flight training

The student must have received and logged instruction in the maneuvers and procedures below, and must have demonstrated satisfactory performance as judged by the instructor endorsing the student for solo.

- Flight preparation procedures and background, including pre-flight inspection, aircraft systems, principles of glider assembly and disassembly, ground handling, towline rigging, launch and tow signals, and release procedures
- Normal and crosswind takeoffs for aerotow, ground launch, or self-launch, as appropriate
- Emergency procedures and response to equipment malfunctions, including towrope break procedures
- Straight glides and shallow, medium, and steep bank turns and spirals in both directions

- Climbs and climbing turns
- Procedures for using lift as appropriate to the training area
- Flight at various airspeeds from cruising to minimum controllable airspeed
- Incipient stall entries and recoveries and full stall entries and recoveries from various flight attitudes
- Ground reference maneuvers
- Descents with and without turns using high-drag and low-drag configurations
- Airport traffic patterns including entry and departure procedures as well as collision avoidance and wake turbulence avoidance
- Normal and crosswind landings
- Slips to a landing

Preparing for your Private Pilot Glider Practical Test

The Private Pilot Glider Practical Test consists of an oral test and a flight test. The contents of each are available to you in the FAA's booklet *Private Glider Pilot Practical Test Standards* (available from the Soaring Society of America — see "For further reading" in the Appendix). This booklet lists the areas to be covered in the oral test, as well as the maneuvers to be accomplished in the flight test. Make it your guide as you and your instructor prepare for your practical test.

Use the following checklist to make sure you have everything ready on the day you take your practical test.

Checkride checklist

- _ Pilot logbook showing required dual instruction, required solo flights, and CFIG recommendation for the checkride
- _ Student pilot certificate
- _ Completed FAA form "Airman's Application for Rating"
- _ Your original private pilot glider written test results
- _ Signed statement from recommending CFIG indicating that you have reviewed areas in which your knowledge was found deficient as indicated by the written test results
- _ Current Sectional map(s), TCA chart (if appropriate), Airport/ Facility Directory, Federal Aviation Regulations
- _ E6B flight computer, 10-key calculator
- _ Flight Service weather briefing, including thermal index
- _ Two-seat glider with all paperwork up to date and available for inspection
- _ Glider flight manual including performance data and weight-and balance-information
- _ Tow pilot, towrope, towplane
- _ Vehicle and towrope for ground handling
- _ Fee for Designated Examiner

Earning your A, B, and C Badges

The A, B, and C Badge Program of the Soaring Society of America honors your accomplishments in learning to fly gliders. The program also prepares you for advanced goal-oriented flying after you have earned your Private Pilot Glider license. The badges are awarded by SSA-authorized instructors. The requirements for each badge are listed below.

A Badge requirements

The recipient of the A Badge must have knowledge of FAR parts 61 and 91, glider familiarization, pre-flight and ground handling, airport familiarization, tow signals and procedures, normal and crosswind takeoffs, straight glides and turns, stalls, and pattern planning and execution. The A Badge is awarded upon your first solo glider flight.

B Badge requirements

The recipient of the B Badge must have had a solo flight of at least 30 minutes after release from a 2,000-foot tow (add 1.5 minutes for each 100 feet of additional tow altitude). The B Badge also can be earned by flying for at least five minutes above tow-release altitude, or by flying for at least five minutes above a previous low point after release.

C Badge requirements

The C Badge requires knowledge of thermal, wave, and ridge soaring techniques. Also required is knowledge of cross-country procedures including flight planning and glider assembly, disassembly, and retrieval.

Flight requirements for the C Badge include an approach with an instructor aboard made without reference to the altimeter, as well as an approach with an instructor aboard during which touchdown and rollout to a stop are completed in a space of less than 500 feet. Finally, the C Badge requires a minimum of two hours of solo flight time in gliders.

Bronze C Badge requirements

To earn the Bronze C Badge you must complete the following requirements under the supervision of an SSA Instructor.

- Earn the A, B, and C Badges
- Perform at least three solo spot landings witnessed by an SSA Instructor

  Landing 1 date: ________ Instructor: ______________

  Landing 2 date: ________ Instructor: ______________

  Landing 3 date: ________ Instructor: ______________

- Log at least two flights that last two hours or longer

  Flight 1 date: ________ Duration: ______________

  Flight 2 date: ________ Duration: ______________

- Log at least 15 solo hours in gliders, including a total of at least 30 solo flights, with at least 10 flights in a single-seat glider

  Date 15 solo hours obtained: ______________

  Date of 30th solo flight: ______________

  Date of 10th flight in single-seat glider: ______________

- Log at least two accuracy landings made without reference to an altimeter (to simulate off-field landings) with an instructor aboard

  Landing 1 date: ________ Instructor: ______________

  Landing 2 date: ________ Instructor: ______________

- Pass the closed-book Bronze Badge Written Exam (administered by an SSA Instructor) with a score of 80 percent or better

  Score: ________ Date: ______________

  SSA Instructor: ______________________________

The Soaring Society of America

The Soaring Society of America (SSA) has served American glider pilots since 1932. I strongly recommend joining, because the SSA:

- Publishes *Soaring* magazine each month
- Promotes the interests of glider pilots in federal airspace hearings
- Promotes the interests of glider pilots in airport preservation
- Publishes dozens of titles of interest to all glider pilots (such as *Soaring Flight Manual* and *The Joy of Soaring*)
- Provides Glider Flight Instructor clinics all around the country
- Provides an annual national forum for all glider pilots, the SSA National Convention
- Provides a national marketplace for new and used gliders and other equipment
- Sanctions all official glider races
- Provides direction to the Soaring Safety Foundation
- Promotes the interests of SSA divisions such as the National Soaring Museum, the Sailplane Homebuilders Association, the 1-26 Association, the Women Soaring Pilots Association, and the Self-Launching Sailplane Pilots Association

Membership in the Society is open to all. I have worked closely with the SSA for years on a number of issues and know first-hand how fortunate we are to have this organization working for us. For membership information, contact:

The Soaring Society of America
P.O. Box 2100
Hobbs, NM 88241
(505) 392–1177

For Further Reading

The Art Of Thermaling . . . Made Easy! (Wander)
Aviation Weather (FAA)
Badge Soaring: The A, B, & C Badges . . . Made Easy! (Wander)
Badge Soaring: The Bronze Badge . . . Made Easy! (Wander)
Badge Soaring: The Silver Badge . . . Made Easy! (Wander)
Breaking The Apron Strings: Soaring Cross Country (Petmecky)
Cross Country Manual For Glider Pilots (Carswell)
Cross Country Soaring (Reichmann)
End Of The Line: Glider Pilot's Aerotow Manual (Shain)
FAA Glider Flying Handbook (FAA)
Glider Polars & Speed-To-Fly . . . Made Easy! (Wander)
Gliding (Piggott)
Safer Soaring . . . Made Easy! (Wander)
Soaring Flight Manual (SSA)
Soaring Accidents That Almost Happened (DuPont)
Towpilot Manual (Compton)
Transition To Single Seat Gliders . . . Made Easy! (Wander)
Understanding The Sky (Pagen)

These books, and many others, are available from:

SOARING BOOKS & SUPPLIES
P.O. Box 17250 Minneapolis, MN 55417
(800) 660-0238 (952) 920-1804
E-mail: Soarbooks@aol.com

Order Online at www.bobwander.com

Progress records

The record on the following pages is for you to record your progress in learning to fly gliders. Your instructor should remove the progress record that is inside the back cover of this book and keep it for his or her records.

In the left-hand date column, write the date a topic or maneuver was introduced. In the right-hand date column, write the date that you achieved the completion standard.

Name

Date topic introduced	Date topic completed	
		Section 1: Pre-flight and ground handling
/ /	/ /	Glider familiarization
/ /	/ /	Pre-flight, ground handling, airport familiarization
/ /	/ /	Flight instruments
/ /	/ /	Glider assembly and disassembly
		Section 2: Basics of free flight
/ /	/ /	Primary flight controls
/ /	/ /	Other flight controls
/ /	/ /	Scanning for traffic
/ /	/ /	Shallow banked turns and short straight glides
/ /	/ /	Medium banked turns
/ /	/ /	Area familiarization, staying upwind
/ /	/ /	Straight-ahead stalls
/ /	/ /	Steep banked turns
/ /	/ /	Slow flight
/ /	/ /	Stalls from coordinated turns
/ /	/ /	Stalls with airbrakes
/ /	/ /	Crabbing during cruising flight
/ /	/ /	Flying in lift
/ /	/ /	Selecting cruise airspeed
		Section 3: Takeoff and aerotow
/ /	/ /	The takeoff checklist
/ /	/ /	Introduction to flying the aerotow
/ /	/ /	Flying the aerotow with stick and rudder
/ /	/ /	Shifting through the wake
/ /	/ /	Boxing the wake

(continued)

Date topic introduced	Date topic completed	
		Section 3: Takeoff and aerotow (continued)
/ /	/ /	Takeoff and aerotow
/ /	/ /	Introduction to premature aerotow release
/ /	/ /	Slack rope on aerotow
/ /	/ /	Premature release and abbreviated pattern
/ /	/ /	Low release and downwind landing
/ /	/ /	Crosswind takeoffs
/ /	/ /	Takeoff without a wing-runner
		Section 4: Pattern and landing
/ /	/ /	The landing checklist
/ /	/ /	Introduction to the landing pattern
/ /	/ /	Normal pattern and landing
/ /	/ /	Introduction to the forward slip
/ /	/ /	Forward slip in the pattern
/ /	/ /	Introduction to the sideslip
/ /	/ /	Crosswind patterns and landings
/ /	/ /	Accuracy landing
/ /	/ /	Landing over an obstacle
		Section 5: Advanced free-flight maneuvers
/ /	/ /	Precision turns
/ /	/ /	Spiral dive entry and recovery
/ /	/ /	Incipient spin entry and recovery
/ /	/ /	Zigzag turns
/ /	/ /	Unusual attitude recovery

Student name

Date topic introduced	Date topic completed	
		Section 1: Pre-flight and ground handling
/ /	/ /	Glider familiarization
/ /	/ /	Pre-flight, ground handling, airport familiarization
/ /	/ /	Flight instruments
/ /	/ /	Glider assembly and disassembly
		Section 2: Basics of free flight
/ /	/ /	Primary flight controls
/ /	/ /	Other flight controls
/ /	/ /	Scanning for traffic
/ /	/ /	Shallow banked turns and short straight glides
/ /	/ /	Medium banked turns
/ /	/ /	Area familiarization, staying upwind
/ /	/ /	Straight-ahead stalls
/ /	/ /	Steep banked turns
/ /	/ /	Slow flight
/ /	/ /	Stalls from coordinated turns
/ /	/ /	Stalls with airbrakes
/ /	/ /	Crabbing during cruising flight
/ /	/ /	Flying in lift
/ /	/ /	Selecting cruise airspeed
		Section 3: Takeoff and aerotow
/ /	/ /	The takeoff checklist
/ /	/ /	Introduction to flying the aerotow
/ /	/ /	Flying the aerotow with stick and rudder
/ /	/ /	Shifting through the wake
/ /	/ /	Boxing the wake

(continued)

Date topic introduced	Date topic completed	
		Section 3: Takeoff and aerotow (continued)
/ /	/ /	Takeoff and aerotow
/ /	/ /	Introduction to premature aerotow release
/ /	/ /	Slack rope on aerotow
/ /	/ /	Premature release and abbreviated pattern
/ /	/ /	Low release and downwind landing
/ /	/ /	Crosswind takeoffs
/ /	/ /	Takeoff without a wing-runner
		Section 4: Pattern and landing
/ /	/ /	The landing checklist
/ /	/ /	Introduction to the landing pattern
/ /	/ /	Normal pattern and landing
/ /	/ /	Introduction to the forward slip
/ /	/ /	Forward slip in the pattern
/ /	/ /	Introduction to the sideslip
/ /	/ /	Crosswind patterns and landings
/ /	/ /	Accuracy landing
/ /	/ /	Landing over an obstacle
		Section 5: Advanced free-flight maneuvers
/ /	/ /	Precision turns
/ /	/ /	Spiral dive entry and recovery
/ /	/ /	Incipient spin entry and recovery
/ /	/ /	Zigzag turns
/ /	/ /	Unusual attitude recovery